ANIMAL RHYME TIME!

RHYME TIME WITH WHALES!

BY LEIGH MCCLURE

Gareth Stevens
PUBLISHING

Please visit our website, www.garethstevens.com. For a free color catalog of all our high-quality books, call toll free 1-800-542-2595 or fax 1-877-542-2596.

Library of Congress Cataloging-in-Publication Data

Names: McClure, Leigh, author.
Title: Rhyme time with whales! / Leigh McClure.
Description: Buffalo, NY : Gareth Stevens, [2024] | Series: Animal rhyme time! | Includes index.
Identifiers: LCCN 2022052667 (print) | LCCN 2022052668 (ebook) | ISBN 9781538285428 (library binding) | ISBN 9781538285411 (paperback) | ISBN 9781538285435 (ebook)
Subjects: LCSH: Whales–Juvenile literature. | Rhyme–Juvenile literature.
Classification: LCC QL737.C4 M3515 2024 (print) | LCC QL737.C4 (ebook) | DDC 599.5–dc23/eng/20221117
LC record available at https://lccn.loc.gov/2022052667
LC ebook record available at https://lccn.loc.gov/2022052668

Published in 2024 by
Gareth Stevens Publishing
2544 Clinton Street
Buffalo, NY 14224

Designer: Claire Wrazin
Editor: Caitie McAneney

Photo credits: Cover Michael Smith ITWP/Shutterstock.com; p. 5 Imagine Earth Photography/Shutterstock.com; p. 7 (top) Andrea Izzotti/Shutterstock.com; p. 7 (bottom) Chase Dekker/Shutterstock.com; p. 9 (main) Drp8/Shutterstock.com; p. 9 (dolphin) Yakubovich Ekaterina/Shutterstock.com; p. 11 Islandjems - Jemma Craig/Shutterstock.com; p. 13 bozmp/Shutterstock.com; p. 15 Hans Viveen/Shutterstock.com; p. 17 (top) Irina No/Shutterstock.com; p. 17 (bottom) slowmotiongli/Shutterstock.com; p. 17 (inset) Timothy S. Allen/Shutterstock.com; p. 19 (top) jo Crebbin/Shutterstock.com; p. 19 (bottom) Apple Pho/Shutterstock.com; p. 19 (arrow) Lyudmyla Ishchenko/Shutterstock.com; p. 21 Andrew Sutton/Shutterstock.com.

Printed in the United States of America

CPSIA compliance information: Batch #CS24GS: For further information contact Gareth Stevens, New York, New York at 1-800-542-2595.

CONTENTS

Boldface words appear in the glossary.

Ocean Friends

Out in the ocean, you might see a tail.

That tail could belong to a big whale!

Whales live across all of Earth's oceans,

swimming in a wavelike **motion**.

They live in water, but they're not fish.

Turn the page for more facts if you wish!

The Whale Family

Cetacea is the name of the whale family.

It includes whales, dolphins, and porpoises, you see.

The two groups are toothed and **baleen**.

The blue whale is the largest baleen ever seen!

Toothed whales include porpoises, dolphins, and more.

Some live far out, some closer to shore.

DOLPHINS

BLUE WHALE

Superpower Sounds

Toothed whales have a special trick. They can find food by making loud clicks. These sounds send out waves that reflect, or bounce off an underwater object. That gives the whale important **information**! This trick is called echolocation.

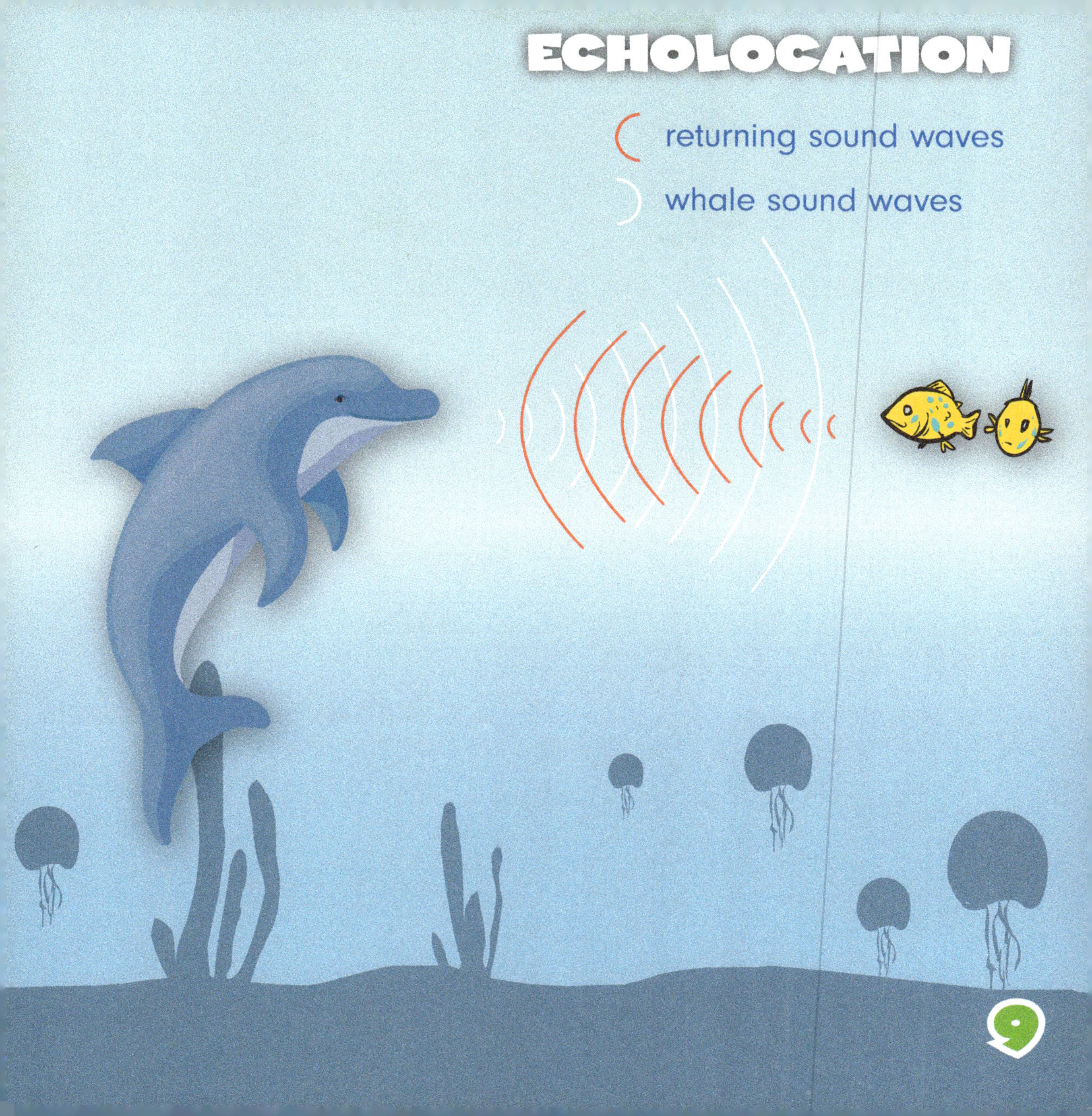
ECHOLOCATION
returning sound waves
whale sound waves

Whale Songs

Did you know baleen whales sing?

Songs tell other whales about things.

Some songs include roars and groans.

Some whales make sighs and **moans**.

Some songs send a warning to others.

Baby whales "whisper" to their mothers.

Whale Bodies

Whales are **mammals** of the sea.

Some **migrate** long ways, wild and free.

Using tails and fins, they swim anywhere.

A blowhole helps whales breathe air.

Blubber is fat underneath their skin.

Blubber keeps their body heat in!

BLOWHOLE

Pint-Sized Porpoises

Looking like dolphins with short **beaks**,
porpoises are small and quite **unique**.
Vaquitas are the smallest whales around.
Harbor porpoises make buzzing sounds.
Porpoises' **dorsal** fins are a bit small.
Some don't have dorsal fins at all!

HARBOR PORPOISE

Daring Dolphins

The ocean has 38 dolphin kinds.

The orca is the largest one you'll find!

These hunters eat squid and fish.

Orcas think seals are a good dish!

Bottlenose dolphins love to play.

They swim around in groups all day!

BOTTLENOSE DOLPHIN

SQUID

ORCA

Baleen Buddies

If you thought orcas were big, just wait!

Baleen whales have huge mouths with comb-like plates.

Mouths wide open, they swim about.

They trap their food and **filter** water out.

Gray whales and humpbacks are in this bunch.

They'll eat tons of **krill** for lunch!

BALEEN
PLATES
GRAY WHALE

KRILL

The Amazing Blue Whale!

Blue whales are the largest
mammal today.

Just how much do these whales weigh?

At 200 tons, these whales are true stars.

Their hearts are as tall as small cars!

With this size, whales need to swim free.

Let's keep oceans a safe place to be!

BLUE WHALE

GLOSSARY

baleen: Comb-like plates found in the mouth of some whales that filter food from the water.

beak: A part of the mouth that sticks out on some animals.

dorsal: On or near the back.

filter: To collect bits from a liquid passing through.

information: Knowledge obtained from study or observation.

krill: A tiny marine creature that looks like a shrimp and is the main food source for some types of whales.

mammal: A warm-blooded animal that has a backbone and hair, breathes air, and feeds milk to its young.

migrate: To move to warmer or colder places for a season.

moan: To make a deep, long sound often due to pain or unhappiness.

motion: Movement.

unique: One of a kind.

FOR MORE INFORMATION

BOOKS

McDougal, Anna. *Humpback Whale Migrations.* New York, NY: Gareth Stevens Publishing, 2023.

Rose, Rachel. *Orca.* Minneapolis, MN: Bearport Publishing, 2021.

WEBSITES

Blue Whale
www.dkfindout.com/us/animals-and-nature/whales-dolphins-and-porpoises/blue-whale/
Discover more fun facts about blue whales.

Bottlenose Dolphins
kids.nationalgeographic.com/animals/mammals/facts/bottlenose-dolphin
Learn more about the bottlenose dolphin—one of the smartest animals on Earth!

Publisher's note to educators and parents: Our editors have carefully reviewed these websites to ensure that they are suitable for students. Many websites change frequently, however, and we cannot guarantee that a site's future contents will continue to meet our high standards of quality and educational value. Be advised that students should be closely supervised whenever they access the internet.

INDEX